The Complete Book of
Birdhouse Construction
for
Woodworkers

Scott D. Campbell

Dover Publications, Inc.
New York

Contents

Introduction 1

I. Construction Basics 3
Construction Materials 3   Entrance Holes 12
Tools of the Trade 5   The Interior 12
Designing the Roof 6   Construction Tips
Cleanouts 8   and Techniques 12
Drainage and Ventilation .. 10

II. The Proper Bird and Birdhouse—The Housing Charts .. 14

III. Detailed Design Suggestions 23
House Finch 24   Downy Woodpecker .. 31
Great Crested Flycatcher ... 25   Wood Duck 33
Purple Martin 25   Bluebird 36
Phoebe 30

IV. Final Placement of the Birdhouse 39
The Easiest Birds to Attract 39
General Site-Selection Suggestions 39
Methods of Hanging and Supporting Houses 41
Inspection 43
Pest Guards 43
When to Place 45
A Final Word 46

Bibliography 46

Copyright © 1984 by Scott D. Campbell.
All rights reserved.

The Complete Book of Birdhouse Construction for Woodworkers is a new work, first published by Dover Publications, Inc., in 1984.

Library of Congress Cataloging in Publication Data

Campbell, Scott D.
  The complete book of birdhouse construction for woodworkers.

  Bibliography: p.
  1. Birdhouses—Design and construction. 2. Woodwork. I. Title.
  QL676.5.C27   1984   690'.89   82-9528
  ISBN-13: 978-0-486-24407-5   AACR2
  ISBN-10: 0-486-24407-5

Manufactured in the United States by Courier Corporation
24407519
www.doverpublications.com

Introduction

One of the greatest joys that any woodworker can have is to create a useful and lasting product that he has designed himself. In the following pages we will explore the basic design features of birdhouses and then progress to actual designs that can be used as presented or adapted to your particular needs. This is only the beginning for the true craftsman, for, after the basic requirements of birds have been met, there is no limit to what your imagination may conceive.

Why are birdhouses necessary? Each year thousands of acres of wildlife habitat are destroyed in order to build shopping centers, subdivisions, and parking lots. Cavity-nesting birds, such as the Eastern bluebird, are affected both by the increased cutting of dead trees for firewood or clearing and the use of nonwood fence posts. In addition, many birds lose nesting sites to more aggressive species such as the starling. The increased populations of cats (and squirrels in protected city areas) make safe nesting sites harder to find. A properly constructed birdhouse can remedy many of these problems entirely and provide immense satisfaction to the owner, in that he is doing his part to help restore the natural balance that man himself has disturbed.

Remember also that while we enjoy both the beauty of birds and their song, they have an important job to do as well. Because birds require a great deal of energy for activities such as flying, reproduction, and warmth in winter, each healthy bird on the average will consume literally many pounds of wild seeds and insects or of rodents and small game each year. Studies of sparrows have shown that they eat one-fourth of an ounce of seeds *per day*.[1] Predatory birds, such as owls and hawks, help keep field mice, snakes, and small game from overpopulation. Imagine the increasing costs to our environment as bird populations decrease. Consider the real costs of increasing herbicide and pesticide use and the hidden costs of more pollution.

You the reader, as a craftsman, hobbyist, or handyman, can enjoy hours of satisfying work in designing and building birdhouses, while at the same time contributing to improving the environment. In an increasingly complicated world, local wildlife-management projects are one of the few areas where direct involvement by an individual on a local scale can not only pay off in locally visible results (increased wildlife activity) but can benefit all Americans as well.

There is also a real commercial need for well-designed birdhouses crafted from durable materials. Investigate any store or hobby center where birdhouses are sold. There you may find some good kits and houses, but most will be of inferior plastics or thin wood frames, or will have other undesirable features. Larger houses (which are more costly to market nationally, owing to freight, packaging, and other costs) are hard to find and generally expensive. For this reason there may be a real market locally for successful designs. Birdhouses are also a great learning experience for young people, either as school or father-son projects. Beginner's kits are often advertised in national catalogs.

Whatever your reasons for using this book, all the designs in this collection can be adapted to the use of simple hand tools. Power tools are great time savers, but please remember to supervise their use by young people very carefully. Only the very proud and the very foolish refuse to use eye and hearing protectors. It is my wish that you will find this book a safe and exciting adventure in creative woodcrafting and wildlife protection.

---

[1]Walter E. Schutz, *How To Attract, House and Feed Birds.* 3rd ed. (New York: Collier Books, 1974), p. 4.

I. Construction Basics

Construction Materials

Wood is the best material for birdhouse construction, for a number of reasons. First, it is easily worked. You can cut, drill, and shape wood in a variety of ways. Second, it is a good insulator against temperature changes and noise. Third, it is widely available in a number of price ranges and in some cases may cost nothing (more on this later). Finally, properly constructed wood birdhouses are durable and can take a variety of different finishes in order to blend with their surroundings.

Wood is available for purchase at lumber and building-supply dealers, conveniently listed in the yellow pages of your phone book under "Lumber." You may purchase stock or standard sizes of plywood or lumber, or if the company offers *millwork* you can have wood cut to your specifications for an additional price. Generally, those companies which advertise to the home handyman or do-it-yourselfer will give better service, but few offer millwork.

What type of wood should you buy? That depends of course on what woods are in stock and also on what characteristics and price range you desire. A durable wood such as cypress, cedar, or redwood will weather very well but is expensive, and cypress is especially hard to nail. A properly treated pine or fir birdhouse will last for years and is easy to construct. Yellow pine may have some pitch or resin pockets in some pieces, which will have to be cleaned and sealed before finishing. White pine is good and the grain is much less noticeable than in yellow pine. Plywood is both strong and convenient for laying out and cutting designs from, but be sure it is graded as "exterior" plywood, which means that waterproof glues were used to construct it. All cut edges of plywood should be sealed against moisture absorption.

What size lumber should be bought? Lumber should be bought in stock sizes to avoid additional millwork costs if possible. Wood thicknesses can be ¼", ⅜", ½", or ¾", as you prefer, though ¼" and ⅜" lumber are more readily available from craft suppliers. Unfortunately, lumber is sold according to its thickness and width before seasoning and surfacing. Thus a board sold as 1" × 2" is actually about ¾" thick by 1½" wide (see Table 1). From the table you can see that to obtain boards at least ¾" thick and 5" wide, you must glue or nail one-by-threes together or buy one-by-sixes and trim if needed. Be sure to inquire about the actual size of the lumber you are buying if it will affect your plans. Stock lumber lengths are in feet: eight, ten, twelve, and greater lengths. Plywood dimensions are exactly as stated. Plywood is available in 4' × 8' panels and some stores offer it in smaller stock sizes, such as 2' × 4' or 2' × 2'.

TABLE 1
Lumber Sizes[2]

Original Size (in inches)	Finished Size (in inches)
1 × 2	¾ × 1½
1 × 3	¾ × 2½
1 × 4	¾ × 3½
1 × 5	¾ × 4½
1 × 6	¾ × 5½
1 × 8	¾ × 7½
1 × 10	¾ × 9¼
1 × 12	¾ × 11¼
2 × 4	1½ × 3½
4 × 4	3½ × 3½

[2] *Lowe's 1974 Buyers Guide* (North Wilkesboro, N.C.: Lowe's Companies, Inc., 1974), p. 11.

Lumber and plywood are also "graded" according to appearance (the smoothness of the surface and the number of knots or waste). Plywood has a letter designation for the quality of its surfaces. For instance, A/C plywood has a superior, smooth surface, designated "A," and a somewhat rougher side, noted as a "C" side. The "A" side will naturally accept a finish better with less effort, and so should face out when the birdhouse is made. Your lumber dealer can help you select the grade of plywood or lumber you need. It is largely a matter of your own preference and the types in stock at the time of your purchase.

What about free wood? Some of the most natural and beautiful birdhouses are made from wood that costs little or nothing at all. Five-inch or larger-diameter logs can be bored or chiseled out to make handsome, sturdy houses. These logs can be found where storms down trees, where trees are cleared for construction, and as driftwood at lake shores. Old packing crates and broken pallets (wooden shipping platforms) are sometimes available at local warehouses. Sturdy, though cheaper and sometimes rough-cut, lumber is used to protect many items that are too expensive or too heavy to be safely shipped in cartons alone. After you have checked the price of lumber at lumber companies you may feel it worthwhile to check with the shipping and receiving departments of larger businesses in your area. A warehouse in this author's area has regularly transported truckloads of damaged pallets to the city dump because the labor and material costs to repair them are not practical.

There is one final type of wood available: rough-sawn wood. Normally wood is cut and then planed to give it a level, smooth surface. Rough-sawn wood, as the name implies, is left in a rough condition for a rustic appearance. Commercial birdhouse builders often groove or otherwise "antique" the wood for that natural or weathered look. Rough wood surfaces are undoubtedly easier also for birds when climbing out of deep birdhouses. This method is most effective with woods like cedar, which age well without painting. Ask your lumberman about obtaining rough-sawn lumber.

Of course wood is not the only material that may be used. Novelty houses have been made from coconut shells, gourds, flowerpots, and various other items, including cement-coated wire mesh. The emphasis in this book, however, will be on the wooden birdhouse.

Tools of the Trade

A brace and bit or small electric drill is required for making entrance, ventilation, and drainage holes. Use ¼" drill bits for ventilation and ⅛" bits for drainage holes if they are required by your design. The most common bit sizes for entrance holes are 1", 1⅛", 1¼", and 1½". An adjustable hole saw for use with an electric drill will also cut many larger-size holes.

The lumber may be cut to size with a handsaw. A keyhole saw or a coping saw is used to cut the big-diameter entrance holes for larger birds. The short and sturdy back saw is helpful in making miter and bevel cuts. If power tools are available, the saber saw or the jig saw, which are adjustable for bevel cuts, will perform all the sawing operations needed, except cutting logs.

The only remaining tools required are a hammer, a standard screwdriver, a ruler, a square for straight lines, and wood chisels and gouges (for working with logs).

Designing the Roof

There are three requirements for a roof: (1) the roof should shed water efficiently to prevent it from draining into the nest box, (2) all exposed joints should be sealed tightly, and (3) the roof overhang should keep blowing rain out of the entrance hole.

Requirement (1) involves sealing the roof surface against water penetration and designing the slope of the roof to direct water where it should go. Requirement (2) involves cutting joints to fit well, and proper sealing of all exposed edges. The amount of overhang in requirement (3) is dependent on how near the roof the entrance hole is and how protected the location is where the house will be placed. Some designers use a two-inch overhang minimum as a rule of thumb.

The simplest roof design is the one-piece (Figure 1). The advantage of this design is that there are no joints to leak and the need for those saw cuts can be eliminated.

The simplest two-piece roof uses a butt joint. One edge is placed at a right angle to the other roof-piece edge. For balance, the board on the low or bottom side of the joint should be shorter than the remaining board by a distance equal to the thickness of the boards used. Interesting effects can be achieved by making one piece much larger than another—particularly if the roof goes from front to back rather than from side to side of the house (Fig. 2).

A variation of the butt joint is the lap joint. A recess is cut into the edge of one board, which allows the edge of the joining piece to fit into it. This design exposes less end grain of the wood and provides more surface area for gluing (Fig. 3).

The most exacting and therefore the most challenging two-piece roof involves beveling or cutting mating roof edges to meet exactly at the middle of the high point of the house and thus match the degree of slope of the roof.

The greater the cutting angle used for a bevel cut, the more sharply the roof will slope. Unfortunately, it is nearly impossible to get a perfect bevel joint without using calibrated

Fig. 1. One-piece Roofs.

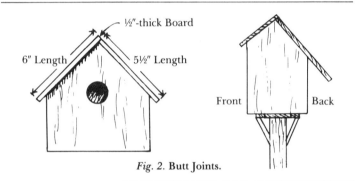

Fig. 2. Butt Joints.

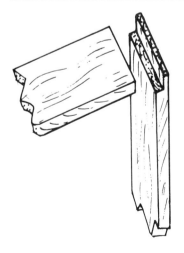

Fig. 3. Lap Joints.

power tools. One method, which requires patience and care, can produce satisfactory results with a minimum of sanding or touch up. Start by acquiring a guide block (made from any scrap of lumber that is several inches wide and thick, such as a two by four) and have it cut to the desired angle at a cabinet shop or lumberyard. Temporarily nail or securely clamp this block to the board to be cut. Lay a saw (preferably a back saw for better control) against the angle cut in the guide block and very slowly cut into the wood beneath the guide. If the saw is kept flat against the guide it will duplicate the guide angle in the wood below (Fig. 4).

Remember that when you cut a bevel in the middle of a board you will have two equal pieces of wood with the same angle. One piece will slope from top to bottom while the edge of the opposite piece slopes from bottom to top. Just turn one piece over and rejoin the boards to form the roof.

At roof peaks, where joints in the wood may leak if not protected, roofs are often overlaid with a strip of tin or aluminum or with roll roofing. A roof made entirely of metal or of asphalt shingles or roll roofing is not advisable, as heat buildup will occur unless the birdhouse is placed in a sheltered, shady area. Adequate ventilation or an insulating wood layer beneath the outer roof shell can also counteract some of the heat. If the heat-accumulating properties of a design are in question, they can be checked by placing a thermometer in the birdhouse. To find scrap roofing material, check construction sites and roofing contractors.

See Fig. 5 for some design suggestions for multisection roofs.

Cleanouts

All types of houses should have a detachable or hinged roof, floor, or side that will allow the house to be cleaned of accumulated debris. This feature reduces the chance of a disease or parasite being transmitted from one breeding season to the next and prepares the house for the new occupant next spring.

One method uses a hinged roof or floor. Hinges may be actual hinge hardware, or devised from a flexible piece of leather, rubber, or tough fabric. Naturally, nonmetallic materials should only be used for roofs, or for sides where the weight of the bird will not be against the hinge, which might fail under stress. Brass screws can be used to secure the cleanout when not in use.

Cleanouts can also pivot on a nail or bar. For this kind of

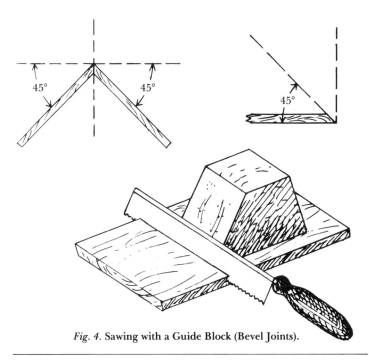

Fig. 4. Sawing with a Guide Block (Bevel Joints).

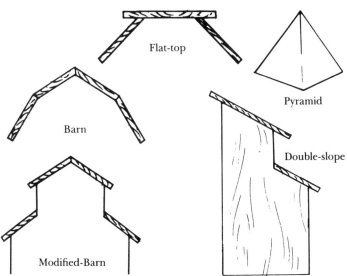

Fig. 5. Multisection Roofs.

cleanout, choose a side or floor that is between two opposing sides of the house. Drill directly opposite holes, slightly smaller than the diameter of the nails to be used, through the opposing sides of the house and into the cleanout to the depth of the nails' lengths. Remove the cleanout and redrill the two opposing holes in the cleanout to the diameter of the nails used. When the house is nailed together the heads of the nails will have a tendency to be held tight in the smaller-diameter holes of the opposing sides, but the part of the nails that extends into the larger holes in the cleanout will allow the cleanout to pivot freely on the nails.

The open-back method involves screwing a birdhouse made without a back to a mounting board which serves as the back. To clean, unscrew the house and leave the mounting board in place. The board may be a smooth wall of a garage under the protecting eave of the garage roof. The outer edge of the birdhouse should be caulked to prevent rain leaks. The backing board should be painted to match or blend with the wall it is attached to.

The sliding-drawer method involves cutting grooves or rabbets in three sides of the bottom of the inside of the house. The fourth side is slotted to allow a floor panel to be pushed in or pulled out. A drawer pull can be attached to the edge and a locking screw can be used on one or more sides to keep the drawer closed when not in use.

To avoid using rabbets, pieces of quarter round or other wood molding or strips of wood can be placed under the drawer to support it and allow it to slide in and out. The drawer itself can be a simple board or a true drawer with front, back, and sides.

Cleanouts can also employ pins, pegs, hooks, catches, movable or stationary cleats, and screws (see Fig. 6).

Drainage and Ventilation

Drainage holes in the bottom of the house may be necessary to keep water from collecting inside during a heavy or extended rainstorm. This problem is most likely to occur if the eaves do not extend far enough beyond the entrance hole to keep out blowing rain or if the roof should develop a leak. A few ⅛" holes in the bottom of the house floor should be sufficient.

Ventilation holes are sometimes also made necessary by the design of the house. If the interior of the house is small and the entrance opening is near the roof, the opening should provide adequate ventilation. If, however, the house is a large

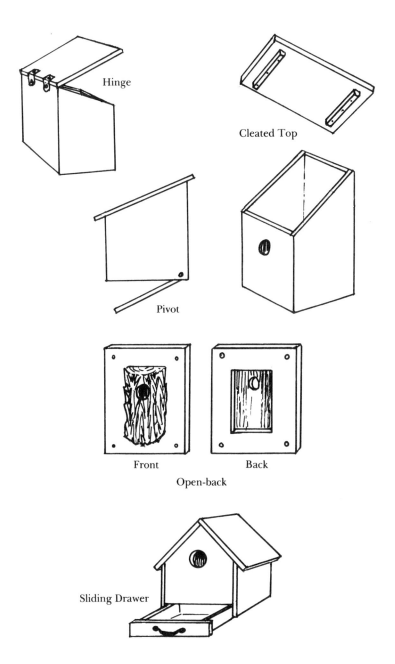

Fig. 6. **Design Suggestions for Cleanouts.**

one with plenty of surface area exposed to the sun or if the entrance hole is well below the peak so that heat can build up near the top of the house, then ventilation holes are advisable. Simply drill two or three ¼" holes above the entrance and near the peak of the roof. Some designers shorten the left and right sides of the house by ¼" so that they don't quite reach the roof and thus allow side-to-side ventilation.

Entrance Holes

The size and shape of the entrance hole is a very critical part of the design. If the hole is too large it either will not attract the bird you wish or may allow predators to get in and rob the nest. Larger and undesired species of birds may appropriate the birdhouse for their own. An example of careful design is the unconventional, elliptical entrance of the wood-duck house or nest box, which is designed to let wood ducks in but keep raccoons out. See Chapter III for detailed design specifications.

If entrance perches are desired, ¼" hardwood dowels can be purchased at hardware stores. The value of these perches to nesting birds is debatable, but they can enhance the appearance of some designs if this is important to the builder.

The Interior

Natural cavities used by birds for nesting have rough-surfaced interiors with plenty of footholds for birds—especially useful to young, inexperienced nestlings exiting the nest. If a birdhouse is both deep and made of smooth, finished lumber, footholds become more important. Hardware cloth, cleats, a narrow incline, or wood dowels glued flat against the surface can artificially roughen the wood for footholds. A series of horizontal grooves cut into the wood below the entrance can help achieve the same purpose.

Construction Tips and Techniques

There are a number of techniques and short cuts that can make birdhouse construction work quicker to complete, longer-lasting, or more beautiful.

First, since most birdhouse designs are simple box shapes, at least two of the sides are usually of the same dimensions. It is therefore possible to clamp or brad two pieces of wood together and make the same cuts on each at the same time.

Secondly, if you use waterproof glue to fasten joints along with nails, the nails will hold the pieces fast until the glue dries.

After the glue has dried, the joints not only are stronger but are sealed better against the rain.

Remember that, when working with woods that split easily, you should predrill nail holes a few sizes smaller than the nails to be used. This is also advisable if softer, nonrusting aluminum nails are used in heavy or dense woods. Aluminum nails are available at hardware stores, building-supply stores, and wherever aluminum-siding products are sold. Brass screws can also be used, will not rust, and are a common hardware-store item.

Some types of bits and woods tend to cause splintering when entrance holes are bored through. For this reason bore from the front side of the entrance through to the interior side and place a block of scrap wood under the exit point area to prevent splinters from forming when the bit exits the wood.

When natural log houses are desired, carefully split or saw the log down the middle of its length. The logs should be thoroughly air-dried beforehand. The inside of the log is then hollowed out using wood chisels or gouges. A series of overlapping holes can also be drilled, but this is time-consuming with hand tools and hard on the bits in heartwood.

Also remember that even plywood that looks smooth will require some light sanding with medium- and fine-grade sandpaper to form a truly smooth finish. In addition, if the method of finishing used on a birdhouse will allow the grain of the wood to show through, it becomes necessary to "balance" the positioning of the grain of the boards. Balancing involves aligning the direction of the grain of lumber so that boards that appear identical to the eye have their grain patterns running in the same direction. Thus, the roof boards would have their grain patterns running in the same direction, as would the left and right sides and the front and back sides. Any of the three groups could have a different grain position but usually the roof boards have their grain running at right angles to that of the four sides.

Use good exterior-grade paints, stains, or varnishes to protect the exterior of the birdhouse. Use dull browns, greens, and grays that will blend with the surroundings. The natural instinct of birds is to seek inconspicuous hideaways in which to rear their young. The only exception to this rule is the attraction of the purple martin to the relative coolness of white birdhouses (white is cooler because it reflects more light than other colors). Do not use finish on the interior of the house—it seems unnatural to the bird and is unnecessary.

II. The Proper Bird and Birdhouse— The Housing Charts

One of the greatest frustrations and at the same time greatest challenges of bird-wildlife protection is the amount of information we still do not have. Some birds, because of their beauty, rarity, peculiar habits, or commercial possibilities, have been extensively studied over long periods of time. Yet habitat-maintenance information for other birds which have less spectacular markings, do not live near human habitations, or have not made an "endangered" list lags behind. This is not entirely due to man's negligence, for there are about 645 species of birds that breed in North America above Mexico.[3] Some species with remote breeding habitats are more likely to be safe from man's encroachments, and a number of subjective criteria, such as an endangered-species list, must be used to allocate limited funds and manpower.

Most birds that have adapted themselves to the use of birdhouses are those known as cavity-nesters. Some cavity-nesters, such as some woodpeckers, will likely never adapt to birdhouses since they prefer to make their own holes and may also never use the same hole twice. These are the original birdhouse builders, for their former holes serve as nest sites for many other species of birds. The only way to help these birds find homes is to allow suitable snags (dead or deteriorating trees) to remain standing in the areas these birds inhabit. Other cavity-nesters can and will adapt to birdhouses. Birdhouse dimensions for some are well known, while statistics on the proper dimensions and use of birdhouses by many cavity-nesters have yet to be established. Table 2 lists 85 species of cavity-nesting birds.

---

[3]Chandler S. Robbins, Bertel Bruun, and Herbert S. Zim, *Birds of North America* (New York: Golden Press, 1966), p. 6.

TABLE 2
Cavity-nesting Birds[4]

Black-bellied whistling duck	Red-bellied woodpecker	Mexican chickadee
Wood duck	Golden-fronted woodpecker	Mountain chickadee
Common goldeneye	Gila woodpecker	Gray-headed chickadee
Barrow's goldeneye	Red-headed woodpecker	Boreal chickadee
Bufflehead	Acorn woodpecker	Chestnut-backed chickadee
Hooded merganser	Lewis' woodpecker	Tufted titmouse
Common merganser	Yellow-bellied sapsucker	Plain titmouse
Turkey vulture	Williamson's sapsucker	Bridled titmouse
Black vulture	Hairy woodpecker	White-breasted nuthatch
Peregrine falcon*	Downy woodpecker	Red-breasted nuthatch
Merlin	Ladder-backed woodpecker	Brown-headed nuthatch
American kestrel	Nuttall's woodpecker	Pygmy nuthatch
Barn owl	Arizona woodpecker	Brown creeper
Screech owl	Red-cockaded woodpecker	House wren
Whiskered owl	White-headed woodpecker	Brown-throated wren
Flammulated owl	Black-backed three-toed woodpecker	Winter wren
Hawk owl	Northern three-toed woodpecker	Bewick's wren
Pygmy owl	Ivory-billed woodpecker*	Carolina wren
Ferruginous owl	Sulphur-bellied flycatcher	Eastern bluebird
Elf owl	Great crested flycatcher	Western bluebird
Barred owl	Wied's crested flycatcher	Mountain bluebird
Spotted owl	Ash-throated flycatcher	Starling
Boreal owl	Olivaceous flycatcher	Crested myna
Saw-whet owl	Western flycatcher	Prothonotary warbler
Chimney swift	Violet-green swallow	Lucy's warbler
Vaux's swift	Tree swallow	House sparrow
Coppery-tailed trogon	Purple martin	European tree sparrow
Common flicker	Black-capped chickadee	
Pileated woodpecker	Carolina chickadee	*Threatened or endangered species.

[4]U.S. Department of Agriculture (Virgil E. Scott, Keith E. Evans, David R. Patton, and Charles P. Stone), *Cavity-Nesting Birds of North American Forests* (Washington: U.S. Government Printing Office, 1977), pp. 5–6.

There are two prerequisites for attracting any particular species of bird to a house design. The bird must be a permanent or summer (nesting-season) resident of your area of the country, and the habitat where the house is placed must match the requirements of the species. The former involves checking the range map of any good field guide, and its importance is obvious. The latter prerequisite, however—habitat requirements—is seldom stressed adequately and is sometimes not even mentioned at all in books on bird housing. It is useless to expect a bird that is normally a woodland nester to choose or even fly through a city site. Birds are no different from any other living creature. They need their preferred food in abundance, protective cover, water, and nesting or shelter sites. Different species of birds find these requirements in different and varied habitats. There are birds of the mountains, birds of open woodlands, pond and river birds, and so on. If a birdhouse is to be functional as well as creatively designed, it and its location must conform to the basic instincts of the bird species to be attracted. A few birds, such as starlings, house wrens, and house sparrows, seem to nest almost anywhere and in any space available. Yet birds only nest where their instinct tells them that their basic and sometimes highly specialized requirements can be met. Imagine the chaos if this were not so. Every bird would have to defend its nesting territory not from a few competing species but from hundreds of species. Smaller, weaker species would be forced into the least desirable environments. Harsher climates would be avoided by all species, and the insects and weeds in those areas would proliferate, while resources of more moderate areas would be depleted.

Fortunately every species has its own niche to fill in the environment. Many exhaustive works on birds and their habitat requirements can be found at the local library. A summary of common or ideal habitats is also included in the house-specification charts at the end of this chapter. Some birds may adapt to environments that are less than ideal, but the more closely the site matches the bird's natural environment the more likely the desired bird will be attracted. For example, wood ducks frequent bottomland hardwood forests and require water nearby for food supplies. Sometimes this habitat is produced artificially by flooding low-lying hardwood forests during nesting season.

Site selection is not always limited to the boundaries of your own backyard. Locate areas suitable to the preferred species near where you live. Perhaps a farmer will allow placement of birdhouses along his fence row. Many businesses have landscaped areas around their premises that are quiet areas frequented only occasionally by gardeners. Local parks may be ideal as well, but place the houses away from heavy traffic areas where vandals will likely be a problem. Churches, nursing homes, hospitals, and other institutions may also give their permission. Be sure to stress that no maintenance will be required by the landowner, and also get the name of the person who gives permission, to prevent misunderstandings later.

The Housing Charts. The specifications charts at the end of this chapter (Table 3) contain lists of names of groups or families of birds that may accept artificial nest platforms, boxes, or houses under proper conditions. Only those species for which specific statistics were available have these dimensions beside their names. If a column or line is blank, the information was not available, and (if you wish to experiment) you are advised to compare the size of the species in question with members of the same family or group for which house dimensions are available. In some cases information was available only on a family rather than individual species. In these cases the dimensions are listed after the "type" name rather than the species names. Pay close attention to habits and habitat information listed in this book and from other sources.

Your choice of species will depend on your own goals. Woodpeckers are exciting house guests but are less likely to use a suitable house than is the house wren. Some houses should be placed high above the ground or near water. The lowly house sparrow is despised by some because of its noisy and aggressive ways but it might be a great first effort for a youngster or a camera buff. Keep in mind the fact that many different species can use certain of the same house dimensions. Undesirable species may have to be evicted before they begin laying eggs to allow the desirable species a chance.

TABLE 3
Nesting Birds and Their Requirements

Types Common Name	Interior Size	Depth	Entrance Size	Entrance Above Floor	Height Above Ground	Habitat
Bluebirds						
Eastern	5″ × 5″	8″	1½″	6″	5–10′	Brushy borders around open areas like pastures—no tall undergrowth.
Western	5″ × 5″	8″	1½″	6″	5–10′	Open forests—particularly of ponderosa pine.
Mountain	—	—	—	—	—	Open forests or forest edges, usually 7,000 to 11,000 feet.
Chickadees						
Black-capped	4″ × 4″	8–10″	1⅛″	6–8″	5–15′	Brushy borders and forests.
Carolina	4″ × 4″	8–10″	1⅛″	6–8″	6–15′	Brushy borders and southeastern forests.
Mexican	—	—	—	—	—	Pine and spruce forests 7,000 to 10,000 feet.
Mountain	—	—	—	—	—	Coniferous forests from 6,000 to 11,000 feet.
Gray-headed	—	—	—	—	—	Broken forests or edges of aspen, willow, and spruce.
Boreal	4″ × 4″	8–10″	1⅛″	6–8″	5–15′	Northern forests of spruce, fir, aspen, and birch.
Chestnut-backed	—	—	—	—	—	Coniferous forests or adjacent woodlands along northwest coast.
Doves						
Rock*	12″ × 12″	9″	(B)	—	10–15′	Farmyards and cities, nesting under bridges and on buildings.

*Captivity-nesting.
(B) One or more sides open.

Nesting Birds and Their Requirements—Continued

Types Common Name	Interior Size	Depth	Entrance Size	Entrance Above Floor	Height Above Ground	Habitat
Ducks†						
Barrow's goldeneye	—	—	—	—	—	Small to medium lakes with open water and aquatic plant and invertebrate food. Nest within 100' of water.
Back-bellied whistling	11" × 11"	22" front –20" rear	5"	—	—	Open woodlands, groves, or thicket borders near water.
Bufflehead*	7" dia.	16"	2⅞"	—	—	Small ponds and lakes in open woodland. Nest close to or over water.
Common goldeneye	12" × 12"	24"	3½" × 4½"	—	—	Lakes and rivers in forested country. Uses open-top or "bucket" cavities.
Common merganser	11" × 11"	34"	4¾"	20"	—	Cool, clear waters of northern or western forests.
Hooded merganser	—	—	3" ellip.	—	—	Wooded, clear-water streams and lakes. Known to use wood-duck houses.
Wood duck	10½" × 10½"	24"	3" ellip.	20"	10–25'	Bottomland hardwood forests near water.
Falcons						
American kestrel	8" × 8"	12–15"	3"	9–12"	10–30'	Brushy borders and open or semi-open country.
Finches						
House	6" × 6"	6"	2"	4"	8–12'	Bottomlands, canyons, suburbs, and ranches.
Flycatchers						
Great crested	6" × 6"	8–10"	2"	6–8"	8–20'	Forest and forest-field edge areas.
Wied's crested	—	—	—	—	—	Deciduous woods and desert saguaros.
Ash-throated	—	—	—	—	—	Western deciduous woods, mesquite, and saguaros.

†Add wood shavings or sawdust to 2- or 3-inch depth.
*Captivity-nesting.

Nesting Birds and Their Requirements—*Continued*

Types Common Name	Interior Size	Depth	Entrance Size	Entrance Above Floor	Height Above Ground	Habitat
Olivaceous	—	—	—	—	—	Dense oak thickets and along canyon streams.
Western	—	—	—	—	—	Moist deciduous and coniferous forests and tall trees near running water.
Nuthatches						
White-breasted	4" × 4"	8–10"	1¼"	6–8"	5–20'	Deciduous woodlands.
Red-breasted	—	—	—	—	—	High elevations of Rocky Mountains and coniferous forests.
Brown-headed	2" × 3"	8–10"	1"	6–8"	5–20'	Burned clearings and areas in southern pine woods and in mixed pine-hardwood forests.
Pygmy	—	—	—	—	—	Western ponderosa pine forests.
Owls†						
Barn	10" × 18"	15–18"	6"	4"	12–18'	Forests, farmyards, marshes, and fields.
Saw-whet	6" × 6"	10–12"	2½"	8–10"	12–20'	Northern forests.
Screech	8" × 8"	12–15"	3"	9–12"	10–30'	Widely spaced tree areas with grassy open spaces—meadow edges, orchards, and forests.
Barred	13" × 15"	16"	8"	—	—	Northern forests, southern swamps, and moist river bottoms.
Phoebes						
Eastern	6" × 6"	6"	(B)	—	8–12'	Common around farm buildings and bridges.
Black	—	—	—	—	—	Western farmyards and along streams.
Say's	—	—	—	—	—	Ranch buildings, bluffs, and cliffs.

†Add wood shavings or sawdust to 2- or 3-inch depth.
(B) One or more sides open.

Nesting Birds and Their Requirements—Continued

Types / Common Name	Interior Size	Depth	Entrance Size	Entrance Above Floor	Height Above Ground	Habitat
Sparrows						
Song	6" × 6"	6"	(A)	—	1–3'	Brushy borders and wood margins.
Starlings						
Starling	6" × 6"	16–18"	2"	14–16"	10–25'	Parks, suburbs, and farms.
Crested myna	—	—	—	—	—	Urban and nearby open fields. Very limited range currently.
Swallows						
Barn	6" × 6"	6"	(B)	—	8–12'	Common near farms.
Purple martin	6" × 6"	6"	2½"	1"	15–20'	Open areas and cutover forests where suitable nest sites are available.
Tree	5" × 5"	6"	1½"	1–5"	10–15'	Usually nest near water and will nest within seven feet of each other.
Violet-green	5" × 5"	6"	1½"	1–5"	10–15'	Open or broken woods or the edges of dense forests.
Thrushes						
American Robin	6" × 8"	8"	(B)	—	6–15'	Open settled country with scattered trees and shrubs.
Titmice						
Plain	4" × 4"	8–10"	1¼"	6–8"	6–15'	Oak and pinyon-juniper woodlands.
Tufted	—	—	—	—	—	Eastern deciduous woodlands.
Bridled	—	—	—	—	—	Prefer chaparral and pinyon-juniper.
Warbler						
Prothonotary	4" × 4"	8"	1½"	5"	4–7'	Swamps and near water in eastern deciduous forests. Nests over or near water.

(A) All sides open.
(B) One or more sides open.

Nesting Birds and Their Requirements—Continued

Types Common Name	Interior Size	Depth	Entrance Size	Entrance Above Floor	Height Above Ground	Habitat
Weaver Finches						
House sparrow	4" × 4"	8–10"	1½"	6–8"	4–12'	Cities, suburbs, and farms near man.
European tree sparrow	4" × 4"	8–10"	1½"	6–8"	4–12'	Suburbs, farmyards, and woodlots in Missouri and Illinois.
Woodpeckers†						
Downy	4" × 4"	8–10"	1¼"	6–8"	6–20'	Open woodland, orchards, farmyards, and urban areas.
Common flicker	7" × 7"	16–18"	2½"	14–16"	6–20'	Near large trees in open woodlands, fields, and meadows.
Golden-fronted	6" × 6"	12–15"	2"	9–12"	12–20'	Mesquite and deciduous woodlands in Texas and Oklahoma.
Hairy	6" × 6"	12–15"	1½"	9–12"	12–20'	Often nests in live trees in open woodlands and forests.
Redheaded	6" × 6"	12–15"	2"	9–12"	12–20'	Open areas—farmyards, field edges, and dead snags in lush ground cover.
Pileated	8" × 8"	12–30"	3–4"	10–12"	12–60'	Extensive mature forest areas.
Red-bellied	6" × 6"	12–14"	2½"	10–12"	12–20'	Common in southeastern forests.
Wrens						
Bewick's	4" × 4"	6–8"	1"	1–6"	6–10'	Farmyards, brushlands, fence rows, and suburbs.
Brown-throated	—	—	—	—	—	Oak forests in desert ranges and in southern Arizona.
Carolina	4" × 4"	6–8"	1⅛"	1–6"	6–10'	Forests with thick underbrush.
House	4" × 4"	6–8"	1"	1–6"	6–10'	Brushy borders and edge habitat.
Winter	4" × 4"	6–8"	1" × 2½"	4–6"	5–10'	Forest brush piles and thick undergrowth.

†Add wood shavings or sawdust to 2- or 3-inch depth.

III. Detailed Design Suggestions

The house designs on the following pages are intended as helpful suggestions and as stimulations to the reader's own creativity. Each design was created with a specific species in mind but is easily altered to accommodate other species of birds. When laying out designs to be cut, allow for some wood to be lost owing to the width of the saw kerf. It is for this reason that no more than one section can be drawn and cut at the same time on the same board unless you are prepared for some irregularities to result.

While any woodworker should stress precision in his work, the builder should not let minor deviations from the plans deter him from using a birdhouse. The dimensions suggested in this book are considered ideal, but of course natural nesting cavities in the wild are far from uniform in size.

The reader will notice that designs are often expressed in fractions. Many beginning woodworkers forget to allow for the width of the boards used when measuring. For instance, if a floor is made of ½" material and is recessed within the house design rather than attached to the bottom (or cut) edges of the sides, then, to be the correct length, the sides must be ½" larger than the design calls for. Or again, suppose a design is made of ¾" material, and the roof is intended only to cover and not extend beyond the left and right sides of the house. The interior space may be 6" × 6", but, if 1½" is not added to the 6" width, the roof will not be wide enough. It should be stressed again that birds can and do accept deviations, and fractions can be ignored in many cases, but be sure of the results before wood is wasted on a useless cut.

House Finch

This design features a simple one-piece roof with a gentle slope, which requires no bevel cuts. Begin assembly by nailing the front and back pieces to the edges of the sides. The floor may also be recessed and nailed along its edges at this time if it is not used as a cleanout. The final step is to align the roof with a 1" overhang on each side, no overhang in back, and a generous 3½" overhang in front.

A simple yet effective way of making a roof cleanout is to use cleats attached to the roof at the inside edges of the house. A flat, heavy object, such as a brick, is then placed on top of the roof to keep high winds from lifting the lid. Scrap quarter round or wood molding is used in this example (Fig. 7) but almost any scrap wood will do. The advantage of the roof cleanout is that it allows examining or photographing the contents of a house without touching them.

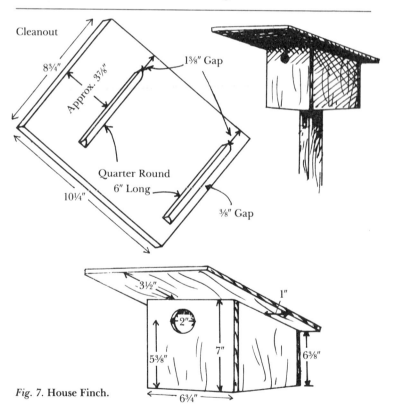

Fig. 7. House Finch.

Construction Details

Material   ⅜" plywood

	No. of Pieces	Dimensions
Floor	1	⅜" × 6" × 6"
Sides	2	⅜" × (7" front tapered to 6⅜" back height) × 6"
Front	1	⅜" × 7" × 6¾"
Rear	1	⅜" × 6⅜" × 6¾"
Roof	1	⅜" × 10¼" × 8¾"

Great Crested Flycatcher

This design features deep-box construction with a peaked roof (Fig. 8). Assembly is begun in two parts. First, nail a 6" scrap piece of wood (in this case a one by two) just below the bevel edge of the underside of one roof section. Now, rest the roof and scrap piece vertically against some support, such as the edge of a workbench, and nail the other roof section to the scrap piece—taking care that the bevel edges of both roof sections mate correctly. Second, nail the sides, floor, front, and back together to form the house. Align the roof on the house and drill one pilot screw hole in each wall near the peak in front and back. Complete assembly by screwing brass screws into the scrap piece to hold the roof in place.

Construction Details

Material   ⅜" plywood, house; ¼" plywood, roof

	No. of Pieces	Dimensions
Floor	1	⅜" × 6" × 6"
Roof	2	¼" × 8" × 8¾"—45-degree bevel
Sides	2	⅜" × 8⅜" × 6"
Front and back	2	⅜" × (11¾" at midpoint tapered to 8⅜" at each outside edge) × 6¾"

Purple Martin

Purple martins, unlike many bird species, live together in close proximity and thus allow the use of multiple-unit houses (Fig. 9). This design features a roof with a generous overhang, and sides extended beyond the apartment openings to achieve greater rain protection. The basic design is for 18 apartments,

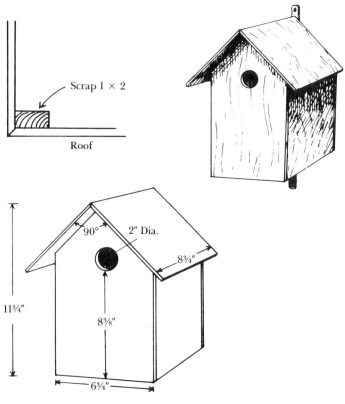

Fig. 8. Great Crested Flycatcher.

but an experienced carpenter can expand this design still further to 20 or 22 apartments by adding dormers to the roof.

Begin construction by forming the first- and second-floor apartment partitions. Cut a 3"-long by ¼"-wide slot in the middle of each short or cross piece. Next cut identical slots at 6" intervals in the central or long partitions. When the slots in the cross pieces are fitted into the slots in the central partitions, the result will be eight equal-sized spaces, each measuring 6" × 6". Center these partitions on the first and second floors and use wire nails and/or glue to secure them.

The remaining steps depend on the cleanout chosen. In the example, the faces or entrance panels of the house are designed to slide out for cleaning. The sides of the house are slotted to hold the faces in place and permit removal from the end. To make the entrances more secure, "shoe" wood molding (quarter round is also good) is applied to support and act as a guide for the sliding panels. The panels will slide more easily if cut ⅛" shorter in height, and this will allow for some swelling of the wood. The molding will hide the gap.

After applying the sides and "attic" floor, construct two apartments for the middle of the end or gable areas. The floor of the attic is used as the floor of each apartment, and cleats may be used if desired to support the walls. The best cleanout is probably a hinged door that is not much larger than the entrance hole itself and is locked in place with a brass screw. Drill ventilation holes in the empty-space areas to the left and right of the attic apartments and near the peak of the roof.

A purple-martin house is the most challenging type of house to build. It represents a big investment in effort, time, and materials. It can't be stressed too much that the house should be thoroughly sealed and painted, and multiple coats are advisable. Asphalt roofing cut into small strips will enhance the beauty and increase the life of the roof. The completed house is also relatively heavy and will require a secure support. White paint is acceptable to purple martins and will make the house cooler.

Construction Details

Material: ¼", ½", ¾" plywood; ¾" × ½" wood molding

	No. of Pieces	Dimensions
Central partitions	2	¼" × 24¾" × 6"
Cross partitions	6	¼" × 12¼" × 6"
First floor	1	¾" × 24¾" × 17¾"
Second floor	1	½" × 24¾" × 17¾"
Attic floor	1	½" × 24¾" × 19½"—45-degree bevel of long sides suggested but not required
Sides	2	½" × (23½" at peak, tapered to 13¾" at each outside edge or 45-degree angle) × 19½"
Entrance panels	4	½" × 25¾" × 5⅞"
Roof	2	½" × 29¾" × 17"—45-degree bevel for roof peak

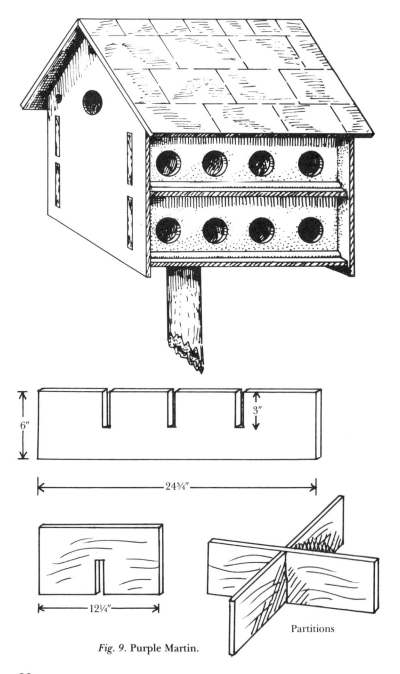

Fig. 9. Purple Martin.

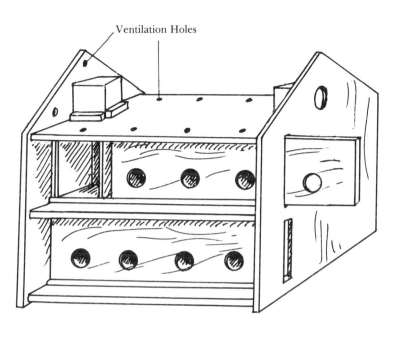

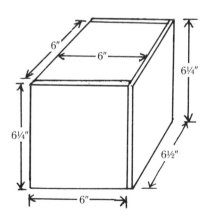

Attic Apartment Detail
One Side and Floor Formed by Attic

Moulding—shoe or quarter round	8	¾" × 24¾" × ½"
Attic apartments:		
Back	2	¼" × 6½" × 6¼"
Sides	4	¼" × 6" × 6¼"
Roof	2	¼" × 6" × 6"

Phoebe

Birds such as robins, barn swallows, and phoebes prefer open-shelf nest sites (Fig. 10). This design features a high roof and three open sides for easy access by the nesting birds. A small enclosure is provided to help secure the nest against winds. Since the nest site is of necessity so open, it is best to place the nest in a sheltered area, such as under the eave of a house or outbuilding.

    Begin construction by attaching the small, raised sides to the floor in a centered position against the back edge. Next nail the floor to a centered cleat and nail the cleat to the backing boards. For extra strength, also be sure to nail the floor to the backing boards by nails driven from the back into the edge of the floor. Nail the remaining cleat to the top of the backing boards to help hold them together and to give the finished work a balanced look. Attach the short braces to the bottom of the floor, and complete construction by nailing on the roof and roof braces.

Construction Details

Material   ½" plywood; 1" × 4" lumber, back; 1" × 2" lumber, cleats*

	No. of Pieces	Dimensions
Floor	1	½" × 9" × 7½"
Roof	1	½" × 11" × 10¼"
Back	2	¾" × 16" × 3½"
Sides	2	½" × 6½" × 1"
	1	½" × 6" × 1"
Cleats	2	¾" × 7" × 1½"
Braces	4	¾" × 6¼" × 1½"

*Actual lumber dimensions are approximately ¾" by 3½" and ¾" by 1½", respectively.

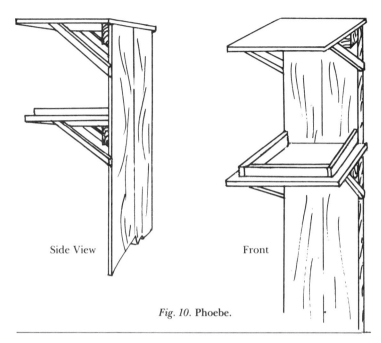

Fig. 10. Phoebe.

Downy Woodpecker

Log birdhouses are the most natural in appearance and perhaps the most beautiful as well. The design illustrated (Fig. 11) was made from a red-cedar driftwood log. It features a hinged top and removable floor plug to facilitate thorough cleaning.

Begin construction by selecting a log that will permit hollowing out a 4"-diameter interior. If the log for this example is approximately 7" in diameter, after trimming the back it will have a 4" hollow and 1"-thick sides. The log should be about 14" tall. The back of the log is sawn along its length to provide a flat surface to fit flush against the mounting board and to allow removal of the interior wood. Driftwood logs often have softer interiors, which are easier to remove. The entrance hole is drilled at a slight downward angle.

The floor plug is composed of a handlebar, a 2½"-diameter plug, and a 2¾" × 2¾" cap. The handlebar allows gripping the plug from below to push it up and out or to help pull it back into position. The plug fits into a circular hole cut in the floor. The square cap on top of the plug extends beyond the floor hole, and since it is placed on the interior side it keeps the plug from falling out. The plug hole can be opened only by pushing

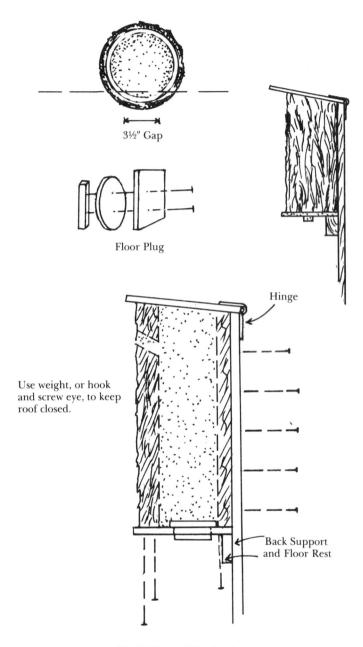

Fig. 11. Downy Woodpecker.

the plug up into the interior of the log; this eliminates the problem of nails or other fasteners giving way under stress.

A hinged top is also used to make inspection easier and to allow placing nesting material inside. One or two inches of clean sawdust, wood chips, or excelsior should be placed in the bottom of the house.

Construction Details

Material: 7"-diameter log; ½" plywood; 1" × 6" lumber*

	No. of Pieces	Dimensions
Floor	1	½" × 7" × 7"
Roof	1	½" × 11" × 11"
Backing board	1	¾" × 28" × 5½"
Floor rest	1	¾" × 2" × 5½"
Floor plug:		
Handlebar	1	½" × ¾" × 2¼"
Plug	1	½" × 2½" diameter
Cap	1	½" × 2¾" × 2¾"

*Actual lumber dimensions are approximately ¾" by 5½".

Wood Duck

The wood duck has drawn the enthusiastic interest of conservationists, who enjoy the beauty and habits of this cavity-nesting bird. While this design (Fig. 12) calls for heavy plywood, it should be mentioned that many professional conservationists recommend unplaned cedar, cypress, or weather-resistant lumber, which will weather well without a finish and last for many nesting seasons. Whatever material you choose should be sturdy and constructed with eightpenny or larger nails (preferably zinc-coated). The house itself should be securely bolted to a support, and that support must have a metal shield or cat guard to keep raccoons, opossums, and other duck predators from robbing the nest.

Begin construction by cutting the entrance hole and tacking the hardware cloth ladder on the interior side below the hole (bend all sharp ladder edges down and under the ladder). Next, assemble the sides around the subfloor, being careful to insure that the ½" floor can move freely in and out of the gap on the shorter side. A ⅛" gap was included in the design both horizontally and vertically to allow a margin of error and for easier movement. Use brass screws, cleat latches, or other

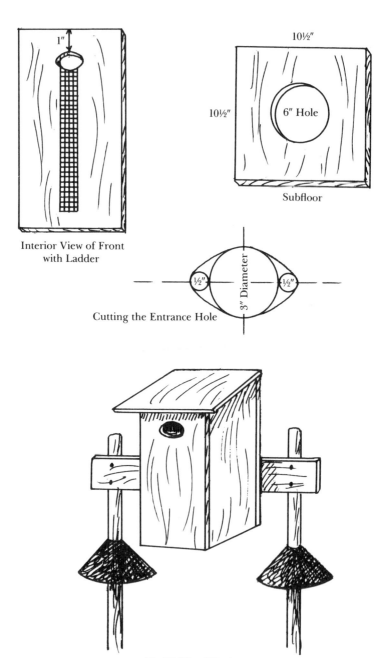

Fig. 12. Wood Duck.

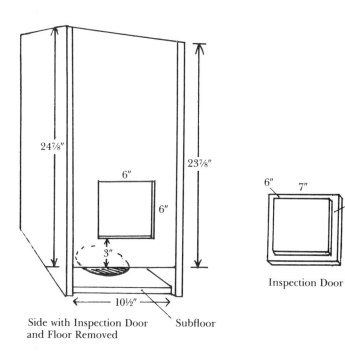

Side with Inspection Door and Floor Removed

Subfloor

Inspection Door

Inspection Door Closed and Floor Partially Pulled Out

suitable arrangements to secure the inspection door in the side. In this design, the inspection door is concealed by an oversized wood cap nailed to the door, which keeps light from entering the house. A cap may also be used for the exposed floor edge. As an added precaution a compressible rubber gasket or suitable weather stripping can be added to the edges of the caps to seal the edges more tightly against rain. Complete assembly by nailing the roof on. A pyramid-shaped roof covered with sheet metal may be helpful if predators are able to jump to the house from nearby trees. Seal and paint the interior as well as exterior areas of the floor to prevent warping, since it will be covered by three inches of sawdust, wood chips, or excelsior.

For more information and design suggestions, write the Superintendent of Documents, U.S. Government Printing Office, Washington, D.C. 20402. Ask for current price information on wildlife leaflet #510, *Nest Boxes for Wood Ducks* (Stock #024-010-00415-1).

Construction Details

Material: ½", ¾" plywood; hardware cloth—¼" mesh

	No. of Pieces	Dimensions
Subfloor	1	¾" × 10½" × 10½" with 6"-diameter hole in center
Floor	1	½" × 11¼" × 10⅜"
Floor cap	1	½" × 10⅜" × 2"
Back	1	¾" × 26¼" × 12"
Front	1	¾" × 25¼" × 12"
Sides	1	¾" × (26¼" back tapered to 25¼" front) × 10½"
	1	¾" × (24⅞" back tapered to 23⅞" front) × 10½"
Roof	1	¾" × 16" × 16"
Inspection door	1	¾" × 6" × 6"
Door cap	1	½" × 7" × 7"
Hardware cloth ladder	1	¼" mesh × 18" × 4"

Bluebird

No selection of birdhouses would be complete without the bluebird house (Fig. 13). Bluebirds are an emotional subject for many people, and the decline of these birds, due to the lack of

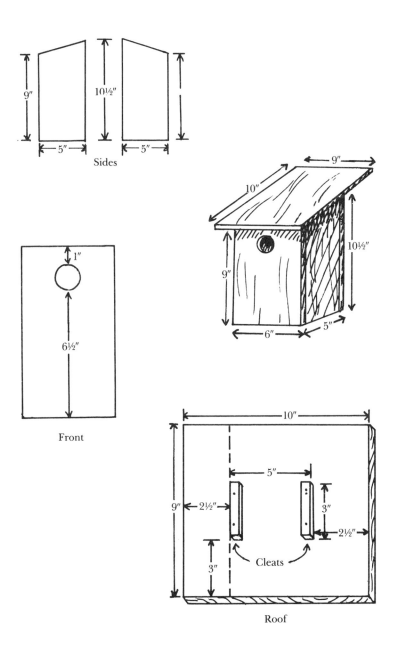

Fig. 13. Bluebird.

nesting sites and the competition for sites with house sparrows and starlings, is no small cause for concern.

Begin construction by assembling the sides, front, and back around the floor. If desired, drainage holes can be drilled in the bottom, and additional ventilation can be added near the peak of the sides. Use cleats under the roof to align the roof and to provide a solid surface to screw the brass cleanout screws into.

Construction Details

Materials   ½" plywood; ¾" × ¾" cleats—lumber is best

	No. of Pieces	Dimensions
Floor	1	½" × 5" × 5"
Sides	2	½" × (10½" in back to 9" in front) × 5"
Back	1	½" × 10½" × 6"
Front	1	½" × 9" × 6"
Roof	1	½" × 10" × 9"
Cleats	2	¾" × 3" × ¾"

IV. Final Placement of the Birdhouse

The Easiest Birds to Attract

A question often arises as to which birds are easiest to attract. The answer depends on where you live and the types of habitats available locally. For this reason there are no easy answers.

If the geographic location is right, if the box design is correct, if the house placement is adequate, and if the available local habitat is satisfactory, the chances of attracting the following species should be good to excellent: in the city—house wren, chickadee, house sparrow, and starling; in the suburbs—house wren, chickadee, tree swallow, violet-green swallow, purple martin, Bewick's wren, tufted titmouse, flicker, house sparrow, and starling; in the country—house wren, chickadee, tree swallow, violet-green swallow, purple martin, bluebird, Bewick's wren, tufted titmouse, flicker, Carolina wren, wood duck, barn swallow, house sparrow, and starling. The difficulties of attracting birds can be compared to those of growing flowers: one person finds it impossible to grow African violets, while the people next door are always giving them away. No two birding enthusiasts are likely to agree on the same list, and the animal world is full of the unexpected. Don't be afraid to experiment!

General Site-Selection Suggestions

After the birdhouse is selected for a general area and built, final placement is all that remains. There are a few simple rules to remember:

> (1) Don't place the house where cats, squirrels, or other bird enemies will be a threat, without supplying adequate protection such as cat guards. A corollary to this rule is "Know your

enemy." Squirrels are great leapers and can jump from nearby trees to a birdhouse. They can also climb small-diameter metal pipe by gripping rather than using their claws to dig in.

(2) Don't place the box near noisy traffic areas—human as well as automotive—or you may discourage more-timid species from using the house and encourage ill-informed tampering by the curious onlooker.

(3) Be sure the house is placed at the correct height.

(4) Be sure the area around the house opening is clear of obstructions which would prevent birds from easily flying to or from the entrance.

(5) The house must be securely fastened to prevent wind damage. Use wire instead of twine to hold a hanging birdhouse, and use a scrap of old garden hose, old tire rubber, or other suitable material under the wire (where it touches the tree limb supporting it) to protect living trees.

Many species of birds prefer edge habitat for nesting: beaches or cliffs near an ocean, edges of clearings in forests, the edge of a pond or river, the forest edge near a meadow, and the shrubby border of a yard or estate. Look for these areas and place a nest platform or house just in front of or just within the edge of one of these border areas. Face the house opening toward the more open habitat: the meadow, the yard, or the pond. In general, partial shade is better than heavy-shade or full-sun areas.

Most bird species establish specific nesting territories and will not let other birds of their species nest within it. The nest territory for a robin may be as little as $3/10$ to $3/4$ of an acre, and for the tree swallow as little as 7 feet, while the white-breasted nuthatch may require 25–30 acres for nesting and foraging, and the black-capped chickadee averages 13.2 acres. The Bewick's wren requires 50–100 yards of space between nest houses, and authorities recommend no less than 100 yards between bluebirds. A suggestion offered by experts for maintaining bluebird houses and which should be adaptable to many species is the use of a bluebird "lane." This concept involves placing bluebird houses at 100-yard intervals along a trail, footpath, or fence row. The builder can walk or drive between the different houses to inspect or maintain them but sufficient distance is established to satisfy the breeding birds and reduce fighting. In backyard situations it is best to use birdhouses designed for several different and noncompetitive species to increase the chances of at least one house being used, yet limiting unproductive fighting.

Methods of Hanging and Supporting Houses (Fig. 14)

When houses are suspended from above, eyebolts are effective. Eyebolts come in a number of different sizes and are common stock items at hardware and department stores. To use, simply drill a hole in the house roof, insert the eyebolt, place a washer over the threaded end to increase the effective area of support, and add and tighten the nut. Another washer placed just below the "eye" of the bolt above the roof will prevent the eye from sinking too deeply into the roof, but is optional. The size of bolt needed will depend on how long the threaded portion has to be to extend through the roof and washer and hold a nut securely. As needed, caulk or seal the area where the bolt enters the top of the roof. Do not confuse the eyebolt with the screw eye. A screw eye has a point at the end and is designed to screw into the wood rather than bolt to it. As the wood of the house ages, and as winds and movement of the occupants increase the stress on the screw eye, it will lose its hold in the wood, and eventually the house will fall. As mentioned before, use only heavy wire, such as that from an unbent coat hanger or other sturdy source. Vinyl-coated clothesline wire is strong and relatively easy to work with. If the wire is looped over a tree, use rubber or heavy vinyl underneath to keep the wire from rubbing or cutting into the bark.

The best method of support is a post, pole, or pipe because the builder can select the exact height and location that he desires and he also has the option of designing the support to pivot or telescope down for inspection and cleaning of the house. Wood posts are usually four by fours, but two by fours can be used where great heights or heavy houses are not required. Use pressure-treated lumber to increase the outdoor life of the supports. Posts can be buried in the ground with $2'' \times 4''$ cleats extending beyond the edges to help keep the post from pulling out. A better method is to tamp gravel under and around the post to allow water to drain away. Still better, set the post or pole on a layer of gravel in a hole $2'$ deep ($3'$ if the post, pole, or house is very heavy or in a deep-frost area) and fill with concrete. The concrete can be mounded and shaped at the top to slope away from the base—thus draining rain water away.

Metal poles and pipes are often used for a purple-martin house because of the height and weight of the house and its exposure to high winds in the open. Various types and diameters of pipes are used by plumbers in water and steam applications, by construction and drilling trades, as conduits in electri-

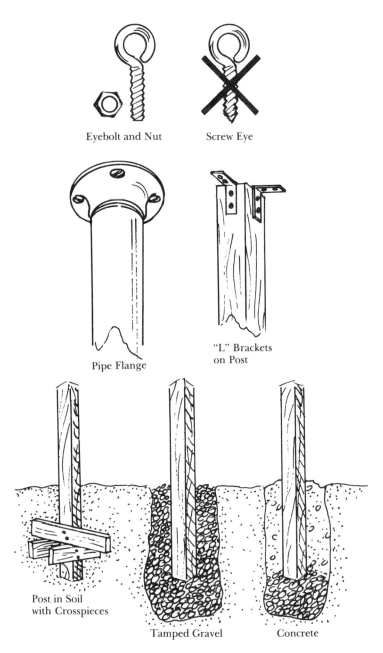

Fig. 14. Methods of Hanging and Supporting Houses.

cal work, and for flagpoles. Beware of thin-walled conduit, which bends and flexes too easily, and pipes and poles which are less than ¾" in diameter, unless height and weight stress will be minimal. Some builders recommend 2" iron pipe for purple-martin houses but this is a matter of individual circumstances and preferences. Pipe flanges can be purchased for pipes of some diameters that will screw to the pipe and that have screw holes for bolting to the house. Flat steel plates that have been predrilled with screw holes can be welded to the top of poles also. Naturally, metal pipes can be quite an investment, so check with salvage companies and scrap-metal dealers before investing in new pipe.

The worst method of hanging a house is to nail it directly to a living tree. Not only is the tree needlessly injured, but a tree large enough to use as a support without serious permanent damage is also too large to enable effective use of cat or squirrel guards.

Inspection

Two ingenious ways have been devised for lowering houses for inspection: the telescoping-pole and the pivot method. The telescoping-pole method uses a small-diameter pipe which fits or nests inside a larger-diameter pipe. A removable bolt or stop keeps the small pole from slipping inside the larger one when fully extended. The pivot method uses a post or pole which is bolted and/or hinged to a ground support (usually another post or a wall). When a locking bolt or bar is removed from the house support, it is then free to pivot on a remaining bolt or hinge attached to the ground support and thus swing down to the ground (Fig. 15). Some variations of the latter method are well illustrated in Walter E. Schutz's book, *How to Attract, House and Feed Birds*.

Pest Guards

To keep squirrels and cats from climbing birdhouse supports, a sheet-metal guard can be added (Fig. 15). You must defeat the leaping as well as the climbing ability of these animals, so place the guard high on the support. Sheet metal can be formed into inverted cones or used as flat sheets. The animal climbs to the guard but can go no higher. The flimsier the sheet the better. Flat sheets can be designed to tip when weight is applied to them by using oversized holes around the support and balancing the sheet on small and/or flexible supports attached to the

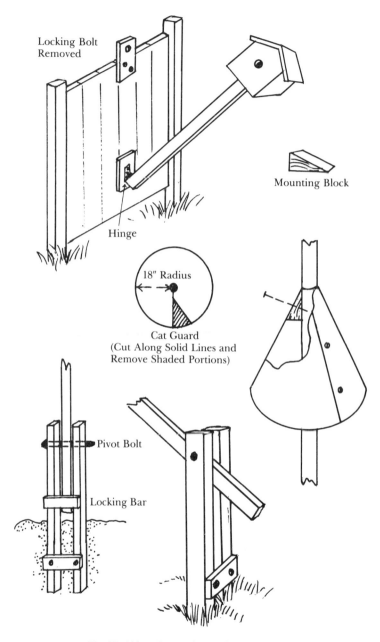

Fig. 15. **Aids to Inspection and Pest Guards.**

pole. A long section of metal can be wrapped directly around wooden poles and tacked in place.

Trapping and extermination of house sparrows and starlings has proven ineffective in the past. If some birds are removed it only serves to create a natural vacuum that will be filled by birds coming in from surrounding areas. Some authorities do recommend placing bluebird houses at low elevations (but no less than three feet) to discourage house sparrows from using them. Nevertheless it must be remembered that, while these introduced birds are not as attractive as some of our native species, they fill an important niche in the difficult city environment. Their actions, even those found disagreeable by some, are purely instinctual, and watching or photographing a mother rearing her young can be as instructional as with any other species. The fact that they are so common around human habitations is a sign of their adaptability and natural resourcefulness. If your view of these birds is less than benign, you may try evicting them from the birdhouses before their eggs are laid, but vigilance must be constant.

Another enemy of nesting birds is, surprisingly to some, the chipmunk. Though chipmunks do not climb as efficiently as some squirrels, they can none the less climb trees and there are reported cases of chipmunks attacking nestlings.

When to Place

Birds arrive for nesting in early spring. Favorable conditions will exist later in the year, the further north you are, so dates vary for different localities. The Carolina chickadee may begin mating as early as February, but April and early May are the busy arrival periods for many bird species migrating to northern areas of the United States. It is also common for some species to have two or three broods, and so the use of nest houses can extend into June or July. Depending on local conditions, the bluebird nesting season can last from mid-March through early August. Robins may build their first nests in evergreens before deciduous trees have developed their full leaf cover. If we also realize that birdhouses should be aged outdoors to allow new paint and other odors to dissipate, it becomes clear that it is never too early to set up a birdhouse. A house erected as late as early summer may have a chance of occupancy, since some species may build new nests for later broods or earlier nests may have been destroyed. But, to be available for the entire breeding season, houses should be in place no later than late winter or very early spring.

A Final Word

After erecting the house, you should inspect it at least once a year and make repairs as necessary during the winter months. Old nests and debris should be removed to prevent parasites from spreading, and rodents or other unwelcome guests should be evicted. Some authorities also recommend removing old nests between broods of the same nesting season to remove old eggs and debris and to keep parasite populations down.

Bibliography

BARBOUR, ROGER W., AND OTHERS. *Kentucky Birds.* Lexington, Ky.: The University Press of Kentucky, 1973.

COLLINS, HENRY H., JR., AND BOYAJIAN, NED R. *Familiar Garden Birds of America.* N.Y.: Harper & Row, 1965.

DAVISON, VERNE E. *Attracting Birds: From the Prairies to the Atlantic.* 2nd ed. N.Y.: Thomas Y. Crowell Company, 1968.

HARRISON, GEORGE H. *The Backyard Bird Watcher.* N.Y.: Simon and Schuster, 1979.

Lowe's 1974 Buyers Guide. North Wilkesboro, N.C.: Lowe's Companies, Inc., 1974.

MCELROY, THOMAS P., JR. *The Habitat Guide to Birding.* N.Y.: Alfred A. Knopf, 1974.

ROBBINS, CHANDLER S.; BRUUN, BERTEL; AND ZIM, HERBERT S. *Birds of North America.* N.Y.: Golden Press, 1966.

SCHUTZ, WALTER E. *How to Attract, House and Feed Birds.* 3rd ed. N.Y.: Collier Books, 1974.

U.S. DEPARTMENT OF AGRICULTURE. *Cavity-Nesting Birds of North American Forests.* Washington, D.C.: U.S. Government Printing Office, 1977.

U.S. DEPARTMENT OF THE INTERIOR. *Fifty Birds of Town and City.* Washington, D.C.: U.S. Government Printing Office, 1978.

U.S. DEPARTMENT OF THE INTERIOR. *Homes for Birds.* Washington, D.C.: U.S. Government Printing Office, 1969.

U.S. DEPARTMENT OF THE INTERIOR. *Nest Boxes for Wood Ducks.* Washington, D.C.: U.S. Government Printing Office, 1976.